TRUE RESCUE

A STORM TOO SOON

TRUE RESCUE
A STORM TOO SOON

A Remarkable True Survival Story in 80-Foot Seas

Michael J. Tougias

with illustrations by

Mark Edward Geyer

Christy Ottaviano Books

Henry Holt and Company • New York

Henry Holt and Company, *Publishers since 1866*
Henry Holt® is a registered trademark of Macmillan Publishing Group, LLC
120 Broadway, New York, NY 10271 • mackids.com

Library of Congress Cataloging-in-Publication Data is available.
ISBN 978-1-250-13756-2 (hardcover)
ISBN 978-1-250-13757-9 (paperback)
Our books may be purchased in bulk for promotional, educational, or business use. Please contact your local bookseller or the Macmillan Corporate and Premium Sales Department at (800) 221-7945 ext. 5442 or by email at MacmillanSpecialMarkets@macmillan.com.
First edition, 2021 / Design by John Daly
Printed in the United States of America by LSC Communications, Harrisonburg, Virginia
1 3 5 7 9 10 8 6 4 2 (hardcover)
1 3 5 7 9 10 8 6 4 2 (paperback)

To my favorite young readers,
Alice, Conner, Rocco, Nellie, and Ben
—M. J. T.

CONTENTS

INTRODUCTION TO SHIP, CREW, AND RESCUERS

SEAN SEAMOUR II CREW

JEAN PIERRE "JP" DE LUTZ (CAPTAIN)

BEN TYE (SAILOR)

RUDY SNEL (SAILOR)

COAST GUARD RESCUE HELICOPTER CREW

NEVADA SMITH (AIRCRAFT COMMANDER)

AARON NELSON (COPILOT)

SCOTT HIGGINS (FLIGHT MECHANIC AND HOIST OPERATOR)

DREW DAZZO (RESCUE SWIMMER)

PART 1

FLORIDA—COUNTDOWN TO THE CROSSING

Sixty-two-year-old Rudy Snel from Canada is standing in the warm Florida sunshine. He is outside an airport waiting to meet Jean Pierre de Lutz, the owner of a sailboat named the *Sean Seamour II.* The 44-foot sailboat will carry Rudy, Jean Pierre, and a third sailor from Florida to France.

Jean Pierre arrives and the two men shake hands, then drive to the sailboat in Jacksonville, Florida. Jean Pierre, who goes by the nickname JP, is excited about the upcoming voyage. He is 57 years old and has been sailing since he was a boy.

When Rudy first sees the *Sean Seamour II*, he is impressed. The sailboat has a center cockpit protected by a hard dodger (rigid windshield), and a single mast rises directly in front of the cockpit. During stormy weather, the cockpit can be completely sheltered with canvas curtains and windows. It's a sleek-looking boat—Rudy thinks it's beautiful.

The third crew member, Ben Tye, emerges from the boat's cabin. JP introduces the 31-year-old sailor to Rudy. Ben is British, with a short, stocky build and a shaved head. He has just sailed from Europe to the United States, and on this trip he will reverse course.

Each man has a different reason for making the voyage. JP wants to have his sailboat near his home in France, Rudy wants the thrill of an Atlantic crossing, and Ben wants to expand on

his sailing experience. For Rudy, the trip will be the culmination of a lifelong dream.

May is the optimal time of year to travel east across the Atlantic, primarily because it puts the sailors ahead of hurricane season. During the next few days, the three men bring food on board and prepare the boat for the voyage. They replace lines, clean equipment, and practice using the pumps that can remove water from the vessel in an emergency.

The men learn about the life raft and the GPIRB: a global position indicating radio beacon, which in an emergency can send a signal to the Coast Guard, pinpointing the boat's location. Also on board is an older emergency beacon from one of JP's earlier boats.

Departure is scheduled for May 1, 2007, but the men have to wait an extra day for some new batteries. These batteries are important for powering the radio, the pumps, and the electronics. However, this one-day delay will have dire consequences. The three sailors don't know it yet, but the *Sean Seamour II* will now be on a collision course with a storm of incredible power.

CHAPTER TWO

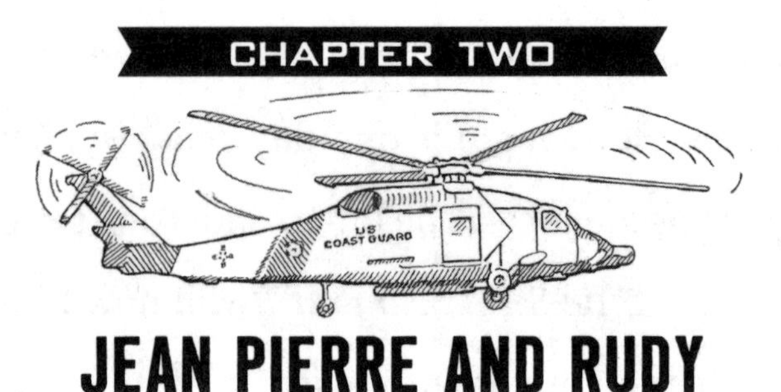

JEAN PIERRE AND RUDY

Before beginning the voyage at six thirty in the morning on May 2, JP checks the marine weather forecast. Everything seems fine.

Using the motor, JP guides the *Sean Seamour II* down the St. Johns River, right through the heart of Jacksonville, Florida. The men reach the open ocean at two in the afternoon.

I'm finally doing it, thinks Rudy as he watches the shore disappear. *We're on our way.*

There's a slight breeze from the southwest, and the men are able to get their sails up. They're at the edge of the Gulf Stream, a current in the ocean that runs northeastward from the tip of Florida along the coast of the United States. The Gulf Stream's current will help propel the *Sean Seamour II* on the first days of the journey.

After admiring the sunset, JP and Ben go to bed. Rudy stands watch to make sure the boat doesn't get too close to any other vessels.

Just before midnight, with a half-moon shining over the water, Rudy has his first thrill of the trip. Peering at the ocean, he sees the outline of a dolphin come through the water. There are about twenty of them swimming along either side of the *Sean Seamour II.* Occasionally, one breaks the surface with acrobatic leaps. Rudy watches, mesmerized. He has never seen dolphins at sea, and this pod wants to race the boat.

Later, JP joins Rudy in the cockpit. JP loves sailing so much, he doesn't want to miss any of it by sleeping. Born in New York City, JP had a difficult childhood. His father, from France, and mother, from Belgium, were not ready to raise a child, so JP spent time in foster care, where he struggled with bullies.

When JP was ten, he was shipped off to France to live with his mother by the ocean. After many stressful years, the sea was soothing to the young JP. When he wasn't in school, he spent much of his free time eagerly watching the fishing boats come and go from the port. Local fishermen eventually noticed JP and taught him everything they knew about the ocean and boats.

Now, many years later, JP viewed his transatlantic crossing aboard the *Sean Seamour II* as a potential first step in sailing around the world.

CHAPTER THREE

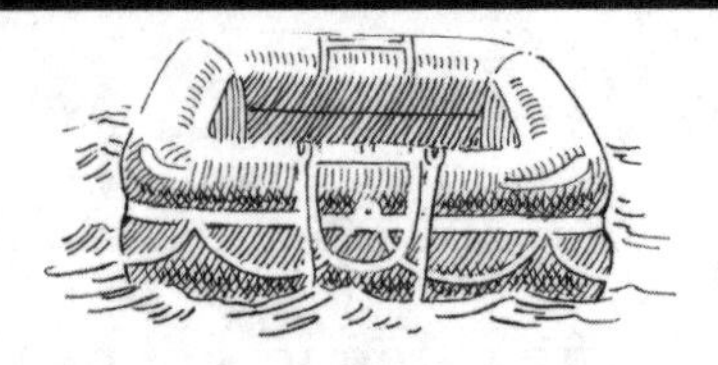

RIDING THE GULF STREAM

May 3 is another fine day, with a brisk breeze from the west. The *Sean Seamour II* is now traveling northeast, riding the Gulf Stream. The ocean's temperature has climbed from 73 degrees Fahrenheit yesterday to 78 today.

The warm waters of the Gulf Stream come from its source in the Gulf of Mexico. The current shoots through the 50-mile-wide Florida Straits, passing between the tip of Florida and Cuba, sending the flow north along Florida's eastern seaboard.

When the Gulf Stream reaches Cape Hatteras, North Carolina, it curves to the northeast, losing a bit of its power. Farther north, the Gulf Stream flows eastward, slowly heading across the Atlantic and beyond the British Isles.

The Gulf Stream can be challenging. Its waters heat the air directly above it, helping to produce a microclimate. When the warm air collides with a cold front, violent thunderstorms erupt.

The Gulf Stream is not the place to be caught in a storm. If winds come out of the north, those winds collide with the water current coming from the south. That combination produces dangerous waves.

On the evening of May 3, Rudy prepares dinner and the sailors enjoy the nice weather by eating in the open air of the cockpit. During dinner, JP mentions that the latest weather report continues to call for favorable conditions. Rudy can't remember when he last felt so relaxed. No phone calls, no television, and no schedules.

May 4 and 5 are also sunny days; with little wind, the crew of the *Sean Seamour II* relies on the boat's motor to propel them forward. On the night of May 5, the captain decides to adjust their course. Some squall lines with rain and wind on the western edge of the Gulf Stream are moving east. The latest weather report also shows two major storm fronts inching closer.

Winds might increase to 25 miles per hour. If bad weather hits, JP wants to be out of the Gulf Stream, where waves can grow larger than the surrounding area because of the current.

For now, JP is still relaxed and enjoying the voyage. He loves his sailboat, loves the sea—but both are about to betray him.

THE GATHERING STORM

At dawn, JP takes over the watch from Rudy. He can't stop thinking about the latest weather report, which shows the low-pressure system to the north of the *Sean Seamour II* gathering strength.

Later in the morning, Ben and Rudy are in the cockpit, thankful for how fast they are moving away from the Gulf Stream. The waves are four feet tall and choppy, but the *Sean Seamour II* slices through them at eight miles per hour.

At 1:00 p.m., the two men notice wispy yellow-brown clouds off to the north.

"What do you make of those clouds?" asks Rudy.

"Rain is coming, and maybe a squall," answers Ben.

Rudy nods. "I don't think I've ever seen clouds that color. They're so menacing."

JP comes up from below, glances toward the stained clouds, and says, "The winds are forecast to veer out of the north to northeast late in the day, and that's when they are supposed to really pick up. We're in for some rough weather, so let's prepare."

The three sailors decide they don't want to get caught with all the sails raised in the event of dangerously extreme winds. They begin lowering the sails but continue advancing toward the east and away from the Gulf Stream.

An hour later, the winds are howling at 50 miles per hour out of the north-northwest. Raindrops, propelled horizontally by the gusts, lash at the three sailors. JP thinks, *This is not good. Nothing is going as planned.*

CHAPTER FIVE

TOO FAR OUT TO SEA

The waves are chaotic and have grown up to 15 feet.

All three men put on their safety harnesses and tether themselves to the boat. The tethers are six-foot-long ropes: if a person in a harness falls overboard, he will not get separated from the vessel and can try to climb back on board.

As the afternoon wears on with no break from the storm, JP considers his options. His original plan to sail north of Bermuda later that night no longer looks feasible. Making a run to Norfolk, Virginia, is out of the question because they would have to cross the dangerous Gulf Stream. They are also much too far out to sea to beat the storm to land. The storm is unstoppable.

JP is soon clocking gusts of wind up to 60 knots, which cause the *Sean Seamour II* to pitch and roll. (One knot is equal to 1.15 miles per hour.) He decides to lay his largest drogue off the stern. The drogue is a wise choice, since it acts as a drag on the boat, counteracting the wind and waves that push the boat forward. Trailed at the end of a long line, the device looks like a parachute with holes on its side. Besides slowing the *Sean Seamour II*, it also helps stabilize the boat. Ben and Rudy assist JP with the drogue.

The three sailors go below, secure the companionway (staircase) hatch behind them, and seal themselves in the belly of the boat. The only thing the men can do now is wait for the winds to blow themselves out.

Daylight gives way to evening, and the storm shows no signs of easing. In fact, it is growing more powerful. The snarling waves are 20 feet tall. Many are steep enough to break, filling the ocean with white water and streaks of foam. Each time a comber—a long, curling wave—hits the hull, the entire boat shudders. A few waves smack so hard that Rudy compares it to being hit by a truck. *Wow,* he thinks, *that felt more like a solid object than a liquid.*

CHAPTER SIX

HELP IS GREATLY NEEDED

Late at night on May 6, the average wave size has increased to an incredible 50 feet. The *Sean Seamour II* is carried up the face of each swell until it teeters at the very top and slips down the back side.

The waves are probing and jabbing the boat for weakness. Rudy can't help but think that the

Sean Seamour II is in the hands of a raging giant, who is shaking and punching the boat, trying to get inside.

✦ ✦ ✦

It's just after midnight, and one of the longest days of the three men's lives is now behind them. *What will May 7 bring?*

JP encourages Ben and Rudy to get some rest. Ben thinks, *This is worse than being held captive. We can't fight back in any way.* It's all up to the boat.

Ben lies down on the port bench, folds his arms across his chest, and closes his eyes. Rudy lurches to the rear cabin and lies down on the bed. JP continues to sit at the chart table. It is now approximately 1:00 a.m.

Five minutes later, there is a thunderous boom. In that second, Rudy is hurled out of his bed, across the cabin, and into the bulkhead, injuring his back. In the salon, Ben, too, is thrown into the air, traveling from the port side toward the chart table, where JP is sitting.

To JP, time has transitioned to slow motion. He can see Ben's arms and legs flailing, propelled sideways through the air. JP instinctively raises his hands to protect himself, and in the next split second, Ben barrels into him.

All this has happened upon the furious impact of a rogue wave. In the next second, the wave sends the *Sean Seamour II* careening over on its starboard side. Loose objects are hurled through the boat like missiles. Canned soft drinks burst like rockets, spraying in all directions. Then, almost as abruptly as the vessel was knocked down, it springs back up, and the unsecured equipment rolls to the opposite side of the cabin.

The men don't know it yet, but the force of the wave is so great that it snaps the drogue trailing behind the boat. Now there is nothing to slow the vessel as the angry seas have their way with it.

CHAPTER SEVEN

ACTIVATING THE EMERGENCY SIGNAL

Rudy and JP half walk, half crawl to Ben and help him up. He is bruised from being launched through the air, but he insists he's okay. JP's gaze sweeps through the galley and salon. He notes that while they are in shambles, there doesn't seem to be any seawater pouring in.

"What happened?" shouts Rudy. There is so

much adrenaline pumping through him that he forgets about the pain in his back.

"It was a knockdown!" hollers JP. "I've got to check the deck, got to see how bad things are!"

JP climbs the companionway, pausing for a second at the top. *Think everything through. Nothing rash.* Carefully, he slides back the hatch. Pellets of ocean spray pepper his face and sting his eyes. The noise from the wind roaring and the waves crashing sends a jolt of fear and urgency down his spine. Amazingly, the deck light is shining, and he can see that the mast is intact.

Then JP's eyes widen as if he's been slapped in the face by an invisible hand. He realizes the entire cockpit windshield and rooftop are gone. The older emergency beacon from his previous vessel, mounted on its cradle in the cockpit, has been swept away.

A wave breaks over the stern, soaking JP and sending water cascading down the companionway. There is nothing more he can do, so he ducks down and seals the hatch.

Ben and Rudy are staring at JP, waiting for him to speak.

"The windshield and roof are gone," he says at last. "I don't know what other damage there is, but we still have the life raft. We're going to need help no matter what happens, so I'm activating the GPIRB. Once the Coast Guard gets the signal, they will call us on the satellite phone. We can let them know what's happening."

JP figures that if the phone won't work, at least the Coast Guard will know there is a boat in trouble and can follow its location as it drifts.

The captain activates the GPIRB, and its

small strobe light blinks, as does the light that indicates the unit is transmitting. Relief settles over the crew of the *Sean Seamour II*. The Coast Guard will know where to find them. The sailors hope it won't be too late.

PART 2

CHAPTER EIGHT

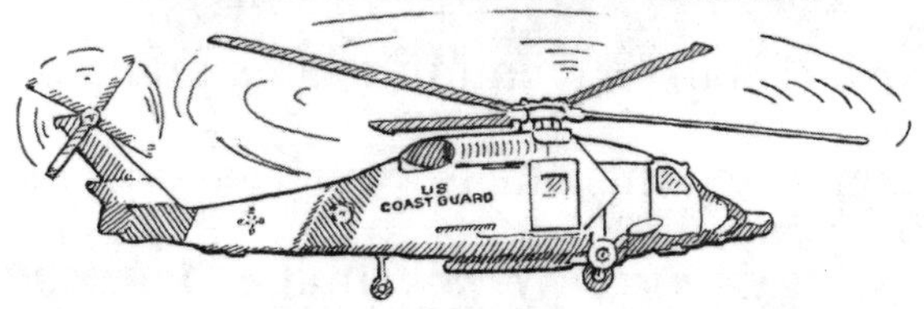

THE WAVE

Although it's night, U.S. Coast Guard search and rescue (SAR) coordinator Brian Avelsgard is busy at the Command Center in Portsmouth, Virginia. Two other coordinators are working alongside him. Each of them has a computer with a dual monitor, a couple of phones, and an assortment of emergency procedure manuals for different scenarios.

Suddenly, a distress notification flashes on their computer screens. It is from the *Sean Seamour II.* Then, just a minute later, another signal comes in. This one is from a sailboat called the *Flying Colours.* "We got another one!" shouts Avelsgard.

One of the coordinators alerts the Coast Guard captain of a C-130 search plane. The C-130 crew should prepare the plane and go to the scene. (Unfortunately, no trace of the *Flying Colours* was ever found.)

On board the *Sean Seamour II*, the men are recovering from the shock of their capsizing. After activating the GPIRB, JP begins to assess the boat's condition. There is some water in the bilge, the lowest section inside the boat. JP is able to pump it out quickly.

The crew has done all they can, so they sit quietly, waiting out the storm. No one can sleep with the wild motion of the boat and the ominous bangs on the hull from breaking waves. As the minutes go by, each man is lost in his own thoughts.

Suddenly the boat starts to roll. It passes the 45-degree mark, and the men hold their breath as their eyes widen in fear and disbelief. For a brief, agonizing second, time stops. They feel that they are carried by some unearthly force as the boat continues to turn and their world goes upside down.

The lights flicker, and a low rumble grows louder from beyond the confines of the cabin. The rolling motion accelerates and defies gravity as objects are hurled from one side of the vessel to the other, crashing upon impact. Rudy and Ben are awkwardly pitched into a tumble as the *Sean Seamour II* does a complete 180-degree roll. JP also starts to fall but is suddenly slammed by the heavy salon table that has snapped off its legs.

The vessel has been hit by a colossal rogue wave that marine experts will later estimate is a minimum of 80 feet tall. It is possible that two waves joined forces to create one mass of energy.

JP struggles to breathe. His head is still above the rising water, but the impact from the falling table has broken seven of his ribs, and he's in excruciating pain. Then he feels the water climb up the back of his neck and head, submerging him in silence.

JP frantically fights to free himself, but he's pinned by the tabletop. *I'm going to drown inside my own boat!*

Rudy is lying on the ceiling in the rising water. It takes him a couple of seconds to process what has happened. A single wall lamp is still on, and the first thing he sees is water shooting through two air vents as if someone is outside with a fire hose. Then he hears shouts: "Help! Help!"

Rudy looks toward the cries. In the dim light, he can see a mess of debris, including the tabletop. The top half of JP's head is barely visible above water.

Ben is also disoriented, but JP's voice snaps him back to reality, and he heads toward the shouting. With Rudy's help, he pushes the table off JP. They pull the captain's head and shoulders out of the water. JP winces and groans. Slowly, he gets to his feet.

Almost two feet of water is sloshing about,

with more gushing in through the ceiling vents below them. Aside from the incoming water, all is quiet.

Somehow, JP manages to ignore the throbbing in his chest. He's got one thing on his mind: the life raft. *Is it still with the boat? If the boat doesn't right itself, we're going to need it.* He looks down where the companionway should be and takes a big breath of air. Searing pain shoots through his ribs. Then he lowers himself into the water and disappears.

Ben and Rudy stare down at the spot where JP dived in. A minute goes by and he does not resurface. The two men look each other directly in the eye. It is their absolute lowest moment. *Did JP get stuck? Is he drowned? Did the waves take him away?* A creeping panic rises between them, spawned by the feeling that they are being buried alive by the sea. And the water is rising.

CHAPTER NINE

THE LIFE RAFT

JP slides open the hatch and swims out, barely able to hold his breath. He needs air, yet he doesn't shoot directly for the surface. Instead, he lets his hands reach their way to the stern of the boat. He sweeps with his arm, feeling for the canister that holds the inflatable life raft.

He can't feel it. It should be just a few inches

behind the arch on the stern of the vessel. His lungs are screaming for air.

Grabbing hold of the toe rail, JP plans to pull himself toward the surface by following the upside-down hull. Suddenly, the boat starts to right itself. It all happens so fast that there is no time to think, and JP holds on as best he can. Incredibly, he lands safely as the boat rolls out of its capsized position.

Coughing, gagging, and desperately sucking in big gulps of air, the captain is exhausted but relieved.

Then he remembers that the life raft is gone, and his spirits plunge as quickly as they rose. He glances under the arch at an empty space where the canister was mounted on the boat. *Without the raft, we are as good as dead. The boat is on borrowed time.*

JP notices that the mast has been cracked, just above deck level, off its base. It is lying to the port side of the boat, almost parallel to the hull. The lower half of the mast is resting on top of the vessel, and its upper half is in the water. The rigging is a complete mess.

JP squints into the ocean beneath where the mast is lying. *The raft!* He can't believe it. The raft is fully inflated and pinned against the sea by the mast.

JP inches toward the raft. It appears to be upside down, and the ballast bags that normally hang beneath the raft and fill with water for stabilization are on the top.

The raft has a tentlike piece of fabric called a canopy that should form a dome above the raft's bottom. But the canopy is torn; it undulates behind the raft with each passing wave. Part of the torn canopy is tangled around the mast, and JP has to cut the canopy completely off. Next, JP notices that there is a foot of space between the mast and the deck. Crawling on his belly, he makes his way underneath this opening.

Lying beneath the mast on his stomach and facing the stern of the boat, JP is only inches away from the raft. He reaches out and touches it with his right foot; he plans to lift the mast and shove it free of the rigging. With the support of his left leg, JP arches his back and boosts the mast. He shoves the raft with his right leg, and the raft bobs free from the mast.

Just as JP begins to lower the mast, an especially large wave breaks directly on him! The mast is pushed down on the left side of his back, crushing him against the deck. He moans in agony. Another three ribs are broken.

Miraculously, he is able to ease out from beneath the mast, where he collapses on the deck, water swirling around his body. The raft is bouncing wildly with each wave. JP gets it on board and secures it.

Sitting in the cockpit, the captain thinks through what should be done next. He wants to wait until the last possible moment to put his crew on the raft. The violence of the waves and the raft's missing canopy mean it will not be an easy ride. *We've got to stay with the boat. We can't leave until I know for sure it is sinking.*

Moving to the companionway, JP can see Rudy and Ben bailing water below in the dim light. He starts down the stairs, ready to tell his crew about the raft. His eyes go to the GPIRB, still in its cradle. He freezes. There is no light from the GPIRB. It's dead.

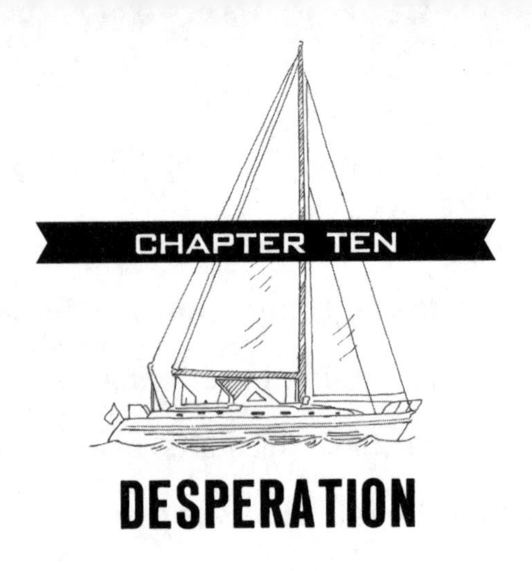

DESPERATION

Luck. JP, Rudy, and Ben are going to need it. The GPIRB is without power, the boat is slowly sinking, and the life raft is damaged.

JP has the yellow GPIRB in his hand, shaking it, turning the switch on and off again and again. There is no light to indicate it's transmitting. *Did it ever send a signal?* he wonders.

JP explains to Ben and Rudy that the life raft's canopy and ballast bags are gone. They'll need to stay with the boat as long as possible.

It is now about four in the morning, and the sailors are exhausted.

JP tries to help Rudy and Ben bail water out of the vessel, but his broken ribs soon force him to stop. He is in pain, but even worse is the feeling that his mind is beginning to cloud over. *Could this be hypothermia?*

Suddenly, at about five thirty, two large waves crash directly onto the boat, sending it careening downward. The men are thrown to the starboard side. Amazingly, no one is injured further.

The boat, however, doesn't fare as well. Green water blasts down the companionway, filling the vessel with another foot of seawater. The level rises above the men's knees.

JP climbs the companionway, and from the cockpit he can see that the bow of his beloved boat is submerged. Holding his broken ribs, he turns and goes back down the steps and shouts, "Let's go! It's time to get out!"

JP sees that his mates are ready. He grabs a yellow ditch bag with survival supplies and lugs it back up the companionway steps. Ben and Rudy follow him up.

When JP crawls into the cockpit, the storm assaults him with shrieking wind and crashing seas. Thankfully, a gray morning light helps them see, and JP is happy to find that the life raft is still secured to the vessel.

JP drops down into it, followed by Ben and then Rudy. They sit in a flat area surrounded by an inflated tube; Rudy compares it to being in a bathtub.

The tether securing the raft to the boat needs to be cut, and Rudy does the job with a knife. By now, the sailboat is nose down in the water. A wave rolls under the raft, pushing them away and blocking the men's view of the sailboat. As the wave passes, the men stare at where the boat should be. It is gone. The ocean has swallowed the *Sean Seamour II* into its dark and unforgiving depths.

CHAPTER ELEVEN

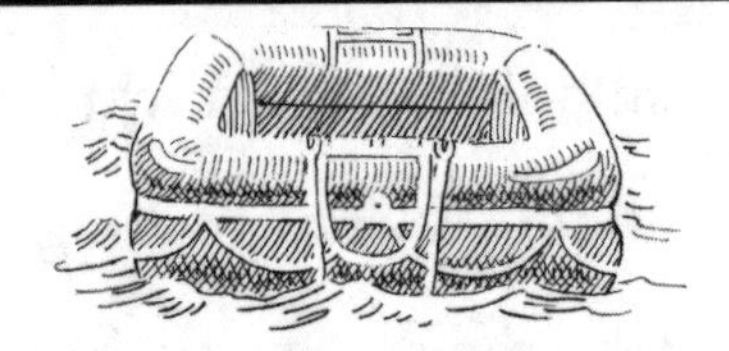

TUMBLING IN CHAOS

JP's beloved boat has sunk and, with it, his connection to the outside world. He stares at the empty spot in the ocean where the boat was just a couple of seconds ago. He tries to be proactive and searches for the bright yellow ditch bag he had filled with supplies. It, too, is gone.

He remembers dropping it in the cockpit and shouting for one of the others to carry it into the raft while he released the tether. But with the shrieking wind and blinding spray, neither Rudy nor Ben heard JP or saw the bag. *It sank with the boat,* he thinks. It was much too heavy to float.

The raft slides wildly with the waves. Whenever it reaches a crest, it's like the men are at the edge of a cliff. They're afraid they will be thrown from the raft. They use the clips at the end of their safety harness tethers to fasten themselves to a lifeline inside the raft.

None of the men speak. All three have arrived at the same conclusion: *No one is coming for us.* Rudy thinks, *Even if we manage to stay in the raft, we aren't going to live long.* He glances at JP and notices the captain's skin color is a sickly gray, as if his blood isn't circulating properly.

After they have been riding on their tiny vessel for about 20 minutes, disaster strikes. An enormous wave slides under the raft, lifting it higher and higher. As the vessel nears the crest, the steep wave can no longer support itself and comes crashing down directly on the raft. It slams into the men with the impact of a train.

The raft is flipped like a pancake, and the men go flying and tumbling. In the trough, Rudy and Ben, still tethered to the raft, realize they are under the overturned vessel. They quickly unclip and kick to the surface. Salt stings their eyes as they search for JP.

Rudy ducks beneath the raft to find him struggling with his tether. Rudy, not yet suffering any real effects of hypothermia, is able to unclip JP and bring him out from under the raft. Rudy hears JP mumbling, and he leans closer.

"Ben, Ben, where's Ben?" says JP.

"He's on the other side of the raft!" shouts Rudy. "He's okay!"

Every part of their bodies is submerged now except their heads. They are expending valuable energy just hanging on to the raft's lifeline, craning their necks so they don't choke on foam or get a mouthful of water. If they stay like this, they won't have to worry about hypothermia—the storm will finish them off by drowning them.

CHAPTER TWELVE
CONVULSIONS

Water pours off Rudy's gray beard. The 62-year-old is being worn down by the effort to hold on to the overturned life raft. So with one final burst of energy he gets the upper half of his body onto the tiny vessel. He heaves his legs up and collapses facedown.

Once Rudy catches his breath, he helps Ben crawl on top. Then the men each grab one of

JP's arms and haul him up. The upside-down raft looks like a trampoline, and the three men have to lie flat on it so they don't slip off. Ben and Rudy position themselves on either side of JP to protect him.

JP's mind is sharp one minute and groggy the next. He's in the fetal position, his arms and legs tucked in tight, and his lips are blue. The wind knifes right through his thin body.

JP knows he is close to the end. He can feel himself convulsing, and the pain from his broken ribs is excruciating. He thinks of his wife, Mayke, in France and wants to tell her he loves her, and that he's sorry for taking this voyage, that he has done his best.

Rudy looks at JP and tries to hold the captain's head in his arms. JP's entire body convulses twice, then it doesn't move at all. *Is he dead*? wonders Rudy.

In a trough between two waves, Ben asks Rudy, a trained pilot, "Do you think a helicopter could fly in these conditions?"

Rudy looks out at the crest of the waves, noting how the wind is spraying off their tops, hurling water sideways through the air. "Probably not."

"Didn't think so," says Ben. He wonders if there is anything he can do to help. Bailing water out of the boat would be futile. The flares are pointless unless someone is searching for them.

The three men are like soldiers on a battlefield, vastly outnumbered and exposed, far from help. They know they will be defeated, but still they defend their little piece of turf: the raft.

PART 3

CHAPTER THIRTEEN

NEED MAX GAS

A telephone ring wakes up helicopter pilot Nevada Smith. He is at home in Camden, North Carolina, about four miles from Air Station Elizabeth City. The caller is the Coast Guard operations duty officer; he tells Nevada they've received an emergency signal from about 220 nautical miles to the southeast. The officer also tells Nevada that a C-130 airplane has already launched and is heading toward the signal.

Nevada says, “I’ll be there in thirty minutes. I’ll need max gas.” That means he’ll need an extra fuel tank on his Jayhawk helicopter to be filled with gasoline. A large man at six foot four and 250 pounds, he is now wide-awake. “I’m on my way.”

Born at the U.S. Naval Air Station in Pensacola, Florida, in 1967, Nevada is the son of a Navy E-2 Hawkeye pilot. It’s not surprising that he wanted to fly from the time he was a little boy.

After college, Nevada enlisted in the Coast Guard and soon attended its extremely demanding flight school. Over the next few years, Nevada flew on dozens of SAR missions, including rescuing victims of Hurricane Katrina in 2005.

Now, arriving at Air Station Elizabeth City, Nevada runs into the Operations Center and puts on his flight suit. The watch captain on duty informs him that he will be the aircraft commander, and his copilot will be Aaron Nelson.

Normally, the commander takes the right seat to monitor the hoist, but Nevada decides that Aaron, an amazing pilot, will be in the right seat. Nevada will take the left.

Although Aaron has been in the Coast Guard only three years, he has more flying experience than Nevada because of his service in the army.

While the two pilots get briefed in the Operations Center, they are joined by Flight Mechanic Scott Higgins and Aviation Survival Technician (rescue swimmer) Drew Dazzo. Nevada is glad to see both men; he now has an expert crew. The four of them race to the helicopter.

The cockpit has countless instruments and switches. One of the most important devices to monitor at the rescue site is the radar altimeter. It measures the distance the helicopter is flying above the ocean. The pilots need to keep the aircraft low enough that Scott can control the rescue cable and get Drew as close as possible to the survivors, but high enough that they don't get hit by a wave. If too much water hits the engines, they will "flame out" and the helicopter will drop like a stone.

CHAPTER FOURTEEN

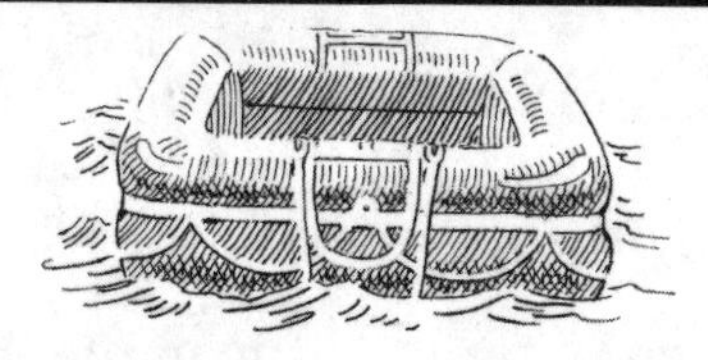

THE JAYHAWK AND THE C-130

As the Jayhawk launches, the commander, Nevada Smith; the copilot, Aaron Nelson; the rescue swimmer, Drew Dazzo; and the flight mechanic, Scott Higgins, settle in for the 250-mile flight. It will take about one hour and 45 minutes. They are headed to the site where the emergency signal came from the *Sean Seamour II.*

The first hundred miles are rough. They're cruising between 500 and 900 feet above the ocean, through rain and wind gusts of 50 miles per hour.

Drew is getting psyched for the rescue, just like an athlete before the big game. He's got the butterflies in his stomach that he always gets when he knows the conditions are bad. It's a feeling he loves; it means his adrenaline is on standby.

Now Drew looks out the window of the chopper, then turns to Scott, who has been doing the same. It's hard to say, but the waves now look to be 40 feet tall. Both men have done all sorts of difficult water rescues, but never in seas this large. And they still have another half hour of flying into the storm.

Farther out over the ocean, the C-130 airplane that launched earlier is near the spot where the emergency call went off. The crew see an object on their radar and tell the pilots. Turning the plane, the pilots steer the aircraft toward the object, hoping they are not too late.

Rudy and Ben see the rescue plane at the same time. They're down in a trough, looking up at a patch of sky, framed on either side by moving walls of green water that seem like mountains. For a brief moment, the plane speeds into view, the orange Coast Guard stripe on its side clearly visible.

JP is conscious and surprises his friends by croaking, "They know we are here . . ." He, too, saw the plane for a second before it streaked behind the towering waves.

"Yes!" shouts Rudy. "They found us!"

Minutes go by that feel like an eternity. Then

the plane reappears. It is still off to the left. Rudy begins to worry. *Maybe they didn't see us. They must have got our first emergency signal, but maybe they don't know exactly where we are.*

Quickly, he yanks one of the flares he had stuck in the pocket of his raincoat right before exiting the boat. Rudy fires the flare. It shoots upward, a bright orange ball of flame heading toward the gray sky. It is designed to go several hundred feet into the air, but it travels about 50 feet, then is caught by the wind and blown sideways into a wave. *Oh no,* thinks Rudy, *the plane missed it.*

But the C-130 lead pilot, Paul Beavis, sees the flare and gets on the radio to SAR Headquarters: "We have spotted a makeshift life raft with survivors. Not sure how many, but we think we saw two people. We have dropped a raft, but it was blown away. We will attempt to keep the survivors in sight for the Jayhawk. Tell the helo crew that visibility is fair to poor, wind gusts just above eighty knots, and the waves are enormous! We have enough fuel to stay on scene for another five hours."

Would that be long enough?

CHAPTER FIFTEEN

LIKE A HOCKEY PUCK ON ICE

The excitement of seeing the C-130 does not last long for Ben. Every 5 or 10 minutes, it zips briefly into view between the passing waves. JP is now unconscious, but Ben and Rudy talk a little. Both conclude that the winds are too much for a helicopter to fly in, and that the C-130 is just trying to keep track of their position.

Time drags on. For the next two hours, the plane circles while the survivors below fight hard to stay with the raft. It seems the raft might have a leak, but it's impossible to know for sure or to fix it.

The C-130 commander, Paul Beavis, gives the Jayhawk helicopter pilot, Nevada Smith, the coordinates and the drift rate of the raft. Both Nevada and his copilot, Aaron, are excited: they won't have to waste precious fuel and time searching.

Aaron estimates that they'll have close to one hour to perform the rescue. *We have time,* Aaron thinks, *and that means everything, with conditions this bad.*

When the helicopter pilots are near where they expect the raft to be, they descend to

300 feet above water. They hover, scanning the seas. Nevada is so shocked by the maelstrom below that he picks up his video camera and films the scene.

Over the radio, Nevada contacts Beavis, saying, "We still can't see them."

Beavis, circling the C-130 above the Jayhawk, answers, "Follow me. We're about to go right over them."

Nevada looks at an oncoming wave. About halfway up the liquid mountain he sees a small black object. *These waves are even bigger than I thought. They must be seventy feet!*

Nevada is back on the radio to Beavis. "Okay, we see it." To Aaron, sitting next to him, he says, "Can you believe this?"

"Unreal," Aaron answers.

"I've never seen waves like this, not even

close. The raft looks like a hockey puck on ice, the way it's moving."

Beavis breaks in. "We will be circling above you at sixteen hundred feet."

Nevada instructs Aaron to a position directly behind the raft, and they hold there to allow Scott and Drew to get ready for the deployment. At this low altitude, the crew has a much better view of the half-submerged black raft, and they can spot three people hanging on to it.

Suddenly, Aaron says, "Did you see that?"

"What?" says Nevada.

"The radar altimeter just went from a hundred to twenty."

"Incredible."

"That's an eighty-foot wave!"

"I'll start calling out the big ones."

Aaron has his priorities. Number one is to

avoid having a wave sneak up on him and clip the underside of the Jayhawk. With waves this big, that's a real possibility, and if that happens, it's all over, for both survivors and aircrew.

CHAPTER SIXTEEN

DEPLOYING THE SWIMMER

Rescue swimmer Drew Dazzo checks his dry suit one last time, making sure the seals are tight. He tells the others he's going off the internal communication system. He removes his headset and gears up with his fins, mask and snorkel, and gloves. Finally, he puts on his neon-yellow rescue helmet.

Scott motions for Drew to come to the door, and the rescue swimmer moves into a sitting position at the edge of the doorframe. Scott hands him the hook and strop, which is a nylon sling that Drew secures under his arms.

Scott is wearing leather gloves so he won't cut his hands on the cable. In his right hand is a pendant attached to a long wire cord that controls the hoist, which is suspended from a steel arm extending from the airframe above the door. Scott also has on a thick belt connected to a safety strap fixed to the back wall of the helicopter cabin. The belt and strap will keep him from falling into the sea.

Nevada calls for deployment.

Scott taps Drew on the chest to signal that he is ready. Drew gives his partner the thumbs-up sign that he, too, is ready to be lowered and begin the rescue.

"Deploying the swimmer."

Using the pendant, Scott pulls up a bit of cable while Drew leans out the doorway until he is in the air. Scott puts a hand on Drew to keep him from being blown back into the airframe. Then he slowly releases the cable and Drew lowers down.

Drew is amazed by the power of the wind pushing him back toward the rear of the aircraft. A strap from his mask is hitting his helmet, making a loud *clack, clack, clack* sound in rapid fire.

Up in the helicopter cabin, Scott is on his stomach, using all his strength to push the cable forward, to keep it away from the tail rotor. Through the radio in his headset, he tells the pilots to move the aircraft forward.

"Good, now hold! Swimmer is in position!"

Drew is within 10 feet of the wave crests but a full 80 feet from the troughs. The raft, sliding with the seas, is about 30 feet away, and he can tell this is as close as Scott can get him.

Drew knows it's time—he's got to do this just right. He feels an electric jolt of energy as a big wave rumbles toward him. When the snarling crest passes five feet below, Drew raises his arms and plunges out of the strop.

Aaron has let the helicopter slide back so that he and Nevada can get eyes on Drew.

"Big wave coming," says Nevada as calmly as possible.

Aaron flies up another 10 feet. The monstrous wave just misses them, and Aaron lets out a sigh of relief.

CHAPTER SEVENTEEN

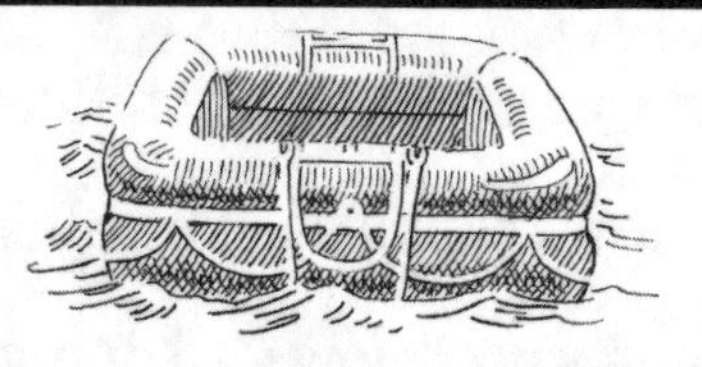

"HE NEEDS TO GO FIRST . . ."

The second Drew hit the water, he started swimming as fast as he could. Now he's rushing to go even faster, scared that he won't catch the raft before a wave breaks and buries him in its fury.

Rudy and Ben are amazed that the rescuer is in the water and actually approaching the raft. Then a wave slides between the raft and the swimmer, blocking their view. When the wave rolls by, the swimmer is at the side of the raft.

Drew puts an elbow on the raft and spits out his snorkel. He takes two deep breaths, then looks at the men and says, "How y'all doing today?"

Ben is so flabbergasted by what he has just seen and heard that all he can do is smile.

Rudy simply says, "You guys are amazing!"

Drew nods and says, "Thank you. Is anyone hurt?"

Ben and Rudy point down at JP, who looks half dead.

Rudy says, "He is. Broken ribs. He needs to go first."

"All right, he's coming with me."

Working inside the cabin, Scott is facing away from the doorway. This is the first time he can't see Drew, and he's concerned. Scott attaches the basket to the hook. He waits for Drew to let him know when he is ready with the first survivor.

Come on, come on, begs Scott, *hurry up.*

"Turn him around!" shouts Drew, referring to JP. "I want his back by the edge of the raft!"

Rudy and Ben slide the unconscious captain around and to the side of the raft. Drew puts his arms under JP's and pulls the captain's back in tight to his chest. He then uses his feet to push against the raft and drags JP out.

JP stirs, and his eyes flutter open. A voice in his ear says, "We're going for a basket ride."

Drew has JP in a cross-chest carry: his right arm is over JP's right shoulder and his hand is under JP's left armpit. The maneuver allows him to tow JP a short distance from the raft.

Looking up at the helicopter, Drew gives the thumbs-up sign that he is ready for the basket.

Scott slides the basket to the doorway. With his left hand, he pushes it outside the cabin while using his right hand to work the pendant, and the basket begins its descent.

Drew sees the basket. He makes his move, plowing through the water with JP under one arm. Just a few feet before his objective, out of the corner of his eye, he sees a wave looming overhead. He grabs a bite of air before he and JP are pummeled by collapsing white water of such force that Drew is stunned and truly frightened. *Whatever happens, don't let go of this man,* he tells himself.

When Drew's head clears the foam, he has swallowed a lot of seawater. Drew treads water and looks at JP, noting his shaking purple lips, yet he's surprised to see that the captain's eyes are open.

The basket is now far away. Drew waits, knowing Scott will guide the pilots back into position, where the basket will be closer. He watches the aircraft maneuvering for another try, Aaron determined not to miss this opportunity.

CHAPTER EIGHTEEN

AS IF SHOT FROM A CANNON

Once again Scott manipulates the cable so that the basket is just 10 feet from Drew and JP. Drew makes his move, dragging the captain with a burst of new energy. He feels the seconds ticking by.

Drew pivots to pull JP up to the side bar of the basket and, continuing that motion, gets the semiconscious captain's body inside.

Not wasting a moment, Drew begins to give the thumbs-up signal to Scott. Just then, a jagged wave passes beneath them. The basket is ripped from Drew's grip. JP and the basket go shooting skyward as if shot from a cannon.

Up above, Scott sees the whole episode play out, holding his breath as the basket jerks violently at the end of the cable. When the basket reaches the helicopter, Scott leans out and uses each of his hands to grab different bails. Then Scott pulls straight back and slides the basket toward the rear of the cabin.

Scott helps JP toward a bench opposite the open doorway, and JP sits. He has a dazed expression, and his eyes stare blankly. But the noise from the aircraft makes JP come out of shock.

As JP's brain kicks into full consciousness, his first thought is: *Where are the others? They can't be gone, they can't be dead.*

✦ ✦ ✦

In the cockpit, Aaron feels a bit of relief that they've rescued one survivor, but now his worry is directed at Drew. Compared with the life raft, Drew is a tiny speck in the ocean, a much smaller visual target. *If we lose sight of him, we will never find him.*

Nevada shouts, "Large wave approaching!"

Aaron quickly nudges up the power to climb out of danger. Then he watches as the colossal wave sweeps up the rescue swimmer and carries him skyward. Drew is lifted so high, he is almost at the same level as Aaron.

Man, I can see his face! thinks Aaron.

Scott is back at his perch by the cabin door. He had a brief view of Drew, but now he's nowhere to be seen. "Back fifty feet," he commands through his headset.

The flight mechanic lies down on his stomach and shimmies his upper body as far out as possible, looking beneath the aircraft. "I see the swimmer! Go back ten more feet."

"Lowering the hook for swimmer," announces Scott.

As the helicopter comes into Drew's field of vision, it's the most beautiful sight he's ever seen. He attaches his hook to the harness and is lifted up toward the aircraft.

Once inside, Drew catches his breath. Then he looks outside the door but can no longer see the raft. *God, I hope we find them.*

BACK INTO THE CHAOS

It takes the pilots three or four minutes of searching the angry gray seas before they locate the raft. Aaron, desperate to save time, speaks to Scott through his headset. "I'm going to try to help you guys by coming down in altitude to see if I can minimize your hoisting distance."

In an amazing display of control, the copilot

lowers the aircraft into the trough, so the crest of the next approaching wave is higher than the helicopter. Aaron has found a rhythm. When the crest is about 40 feet away, he increases altitude, lets the wave pass, and then lowers the vehicle back into the next trough.

Scott manages to position Drew 15 feet downwind from the raft. Drew releases and barely has to swim; the raft is carried straight to him on a wave. Drew is thankful beyond words; he's dog-tired, and this allows him to conserve a little energy.

The swimmer puts an arm on the raft and spits out his snorkel. "Who's next?"

Ben points to Rudy.

Drew explains he will drag Rudy to the basket.

Rudy watched the procedure with JP, so he knows what to do. He turns to Ben and says that he will see him up in the plane. Then he rolls out of the raft and lets Drew pull him a few feet away from the little vessel that has kept them alive against all odds. The basket is lowered within seconds, and Drew helps him inside it.

Once Rudy's safely in the basket, hypothermia, stress, and exhaustion envelop him in bone-crushing stiffness.

Scott is waiting at the helicopter door, working the pendant. When the basket arrives, he hauls it in. He's concerned about this survivor. The gray-bearded man struggles to get out of the basket, and when he does, he lands on all fours. Seawater pours out of his foul-weather gear. Then he collapses on the cabin deck, rolling over on his back.

Scott is thinking heart attack. "Are you okay?" he shouts.

"Yes," gasps Rudy. "I'll be all right."

Scott is thrilled to hear the survivor talking, making sense. He knows that precious seconds are flying by. He's feeling the strain of not knowing where Drew is, and the clock is ticking down to "bingo" time, when they'll run low on fuel and must leave.

Scott sees that Rudy cannot raise himself, and with a surge of pure power, the flight mechanic crouches, grabs Rudy under his right arm, and pulls him up and onto the bench next to JP.

With tears in his eyes, JP hugs Rudy. Both men never expected to be alive, never dreamed a rescue was possible.

CHAPTER TWENTY

A FRAYED CABLE

While Rudy was being hoisted, the life raft carrying the last survivor drifted far from the scene. Ben is alone, and four long minutes have gone by without a glimpse of the helicopter.

Drew is more than a couple hundred yards away from Ben, and although he is not with the raft, he's in a safer position. The helicopter is overhead, and the pilots aren't taking their eyes off him.

Scott has moved the basket to a secure location in the cabin and attached the strop to the hook. "I'm lowering the strop," he announces.

When the strop reaches Drew, he puts it under his arms. Scott retracts some cable. The rescue swimmer is now dangling at the end of the cable just above the wave tops. The pilots slowly move the aircraft, searching for the life raft and the last survivor.

Drew sees the raft about the same time the rest of his crew does, and he shakes his head in wonder. *It looks like a surfboard! Scott will have to get me really close.*

As soon as a wave crest passes beneath the raft, Drew can feel himself being lowered. He's not sure whether the helicopter is descending or if cable is being fed out. Either way, it works.

Drew is only 15 feet above the waves. *This is as close as they're going to get me.* Up come his arms, and he falls from the sling feetfirst. As soon as he hits the water, he's sprinting to the raft.

A breaking sea transports him, and when Drew emerges from the turbulence, he's not sure where the raft is. He swivels around. *There it is! It's getting by me!* He puts his head down and kicks with everything he's got, grabbing the raft before it's gone.

Ben reaches out and holds on to Drew's harness. Drew, gasping for air, spits out the snorkel and holds on to the raft for a couple of seconds, catching his breath. "Okay!" he shouts to Ben. "Come out backward."

Drew locks one arm around Ben and tows him a few feet away, using his free arm to motion to Scott for the basket. Next, Drew secures Ben in the basket and gives the thumbs-up sign. Ben is lifted into the sky, leaving the rescue swimmer alone in the ocean.

Scott uses his left arm to push the cable as far away from the aircraft as possible while it slowly retracts through his gloved hand.

He suddenly stops the hoist. Frayed strands of cable have sliced through his glove, just missing his skin. With the jerking motion of the helicopter, it's hard to see how many strands have broken free.

Scott knows the cable won't snap, but it might get balled up and stuck in the hoist. He prays that the cable will survive this hoist, the final one for Drew.

CHAPTER TWENTY-ONE

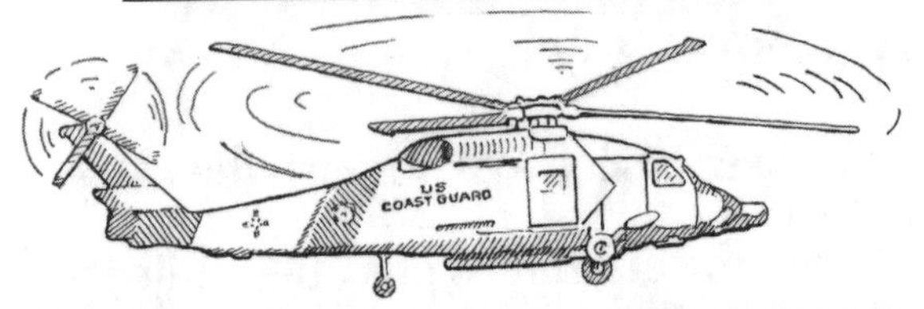

STAY CALM, NO MISTAKES

Scott guides Ben into the aircraft. Ben gets a leg out of the basket and rolls onto the floor. JP and Rudy reach out, tapping his head, patting his shoulder repeatedly, as if to make sure it's really him and not just a dream.

Roller after giant roller slides under the rescue swimmer. Drew is vomiting violently, his body trying to empty out all the seawater it's ingested. He's been exerting himself nonstop for almost half an hour, and he's spent. Over and over he tells himself, *Just hang on for a couple more minutes.*

Then—disaster. A giant wave breaks directly on the swimmer, driving him down while his mouth is open. He can feel himself ingesting more seawater, coughing and choking as he tumbles inside the jaws of the wave. He claws at the water, trying to propel his body in the direction he thinks is the surface. *Air, air,* his brain screams.

Aaron watches in horror as the house-size wave avalanches onto the swimmer. White water swirls in all directions. There is no sign of Drew.

Aaron holds his breath, craning forward to spot Drew's neon helmet. *Come on, come on,* a desperate Aaron urges, *you can do it,* trying to will his rescuer to the surface.

Drew breaks through the churning water. He gulps in a mix of foam and air, frantically waving his arm back and forth over his head—the emergency signal. He's got to get out of the water, knowing he won't survive another wave like that.

Aaron sees Drew first. "I'm getting the emergency signal!" shouts the copilot. "I'm moving closer."

Scott, who is disconnecting the basket from the hook, can't believe his ears. In all his many rescues, this has never happened, and Drew is the toughest swimmer he's ever worked with. Adrenaline surges into every muscle of the flight mechanic's body, and he tells himself, *Stay calm, no mistakes.*

Scott starts lowering the hook, relieved when the frayed cable functions normally.

Drew is treading water, watching the hook descend, hoping he won't have to swim more than a couple of strokes to it, not knowing for sure what his body is capable of anymore. With relief, he sees how Scott and the pilots have worked the helicopter so that the hook is within five feet. He manages a couple kicks of his flipper, grabbing the hook with his right hand. He attaches it to the ring in the front of his harness, located near the bottom of his rib cage.

A second later, as Drew struggles with the straps, a wave crest passes beneath him. There is almost no slack in the cable, so he doesn't slide down the back of the wave. Instead, without the water to support his body, he is yanked so violently that his head and back jerk backward and he feels like he has been snapped in two.

A sharp, white-hot pain shoots up Drew's back, and he moans in agony. The cable is

bringing him higher, but instead of coming up in a seated position, he has his belly at the highest point and his back arched awkwardly. He flails his arms, trying to grab the front of his harness or the hook—anything to help him sit up before his back breaks.

CHAPTER TWENTY-TWO

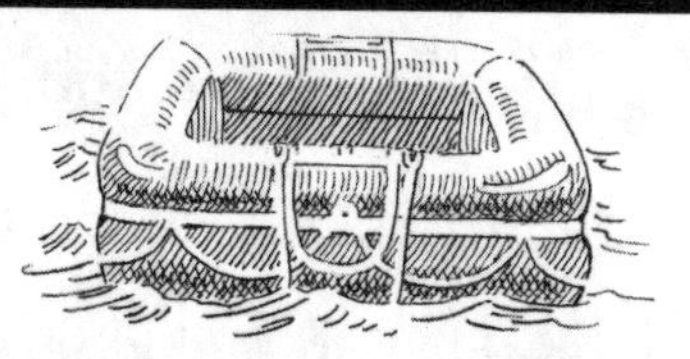

THE LAST MAN

The helicopter actually shudders from the weight of Drew's fall.

"Oh no!" cries Scott. Then a five-second pause. "I might have broken his back."

Aaron hears this comment through his headset, and a cold chill races up his spine. *Did I just hear that right?* He glances at Nevada, whose face is filled with concern. The pilots can't stand the tension, because there is nothing they can do.

Scott's right hand is pressing the pendant while the retracting cable slides through his gloved left hand. He can feel where the frayed cable strands are sticking out. He holds his breath, hoping the hoist mechanism doesn't jam when the loose ends are pulled through.

Scott feels like everything is going all wrong and they are on the verge of disaster. The lift seems to be happening in slow motion. Scott is trying to see Drew's face, wondering if he is conscious. A million thoughts race through Scott's mind, but the main one is: *If his back is broken, I'll never forgive myself.*

Scott gets the swimmer to the door and hauls him in. Drew collapses on the floor. Scott shouts, "Are you okay? Are you okay?"

Drew can't talk but tries to nod. He's light-headed and woozy as he moves toward the closed cabin door. He takes a sip of water and immediately knows he's going to be sick again. He can feel the bile rising, and not wanting to stink up the cabin, he grabs one of his flippers and pukes in it. Then he opens a small six-inch hatch in the cabin door and dumps the contents.

When he turns around, the survivors are staring at him again, and Scott is shaking his head. Drew manages a weak smile and then tries to lie down, but the contractions in his back make straightening out impossible, so he sits facing the survivors, wondering if he's going to pass out.

Scott moves over to Drew and starts apologizing for not being able to release enough cable.

Drew stops him in midsentence. "Listen," says the rescue swimmer, "you got me out. That's all that matters."

Nevada Smith is so proud of his crew that

he needs a moment before he can speak. He composes himself and updates the C-130 that is escorting them back. Nevada then takes the controls from Aaron, who slumps in his seat, exhausted. The three survivors are quiet in the back of the helicopter. Each is lost in his own thoughts, but they all know how lucky they are to be alive. And for that, they are thankful.

EPILOGUE

Rather than fly back to Elizabeth City, Nevada decided to save time by going to Marine Corps Air Station Cherry Point. Once the aircraft landed, Drew and the three sailors were taken to the hospital. All were treated and released at the end of the day.

One simple but crucial fact is that everyone involved—both rescuers and survivors—can look back with pride on how they performed under incredible stress. The three survivors, who had known one another for only a few days, worked as a team. They never once argued. Even when they thought rescue would never come, something made them battle on. Just as remarkable, they never let their fear turn into panic, but instead handled themselves with bravery and dignity. The same can be said for the rescuers. This mission was fraught with so many challenges, it's truly amazing that each man focused on his job and kept making the right decisions time and time again. Everyone simply did their best and never gave up. Together, they accomplished what appeared to be impossible.

AUTHOR'S NOTE

I'd like to thank JP, Ben, and Rudy for sharing their harrowing ordeal with me. I know it wasn't easy to relive those terrifying moments at sea. Many people might have given up or panicked under such pressure, but these three men fought on.

I'd also like to thank the rescuers for being patient with me in recounting the steps they took to save the lives of the sailing crew. Scott, Nevada, Aaron, and Drew: you are true heroes. You have inspired me and countless others with your courage and skill.

PHOTO GALLERY

[Courtesy Jean Pierre de Lutz]

Jean Pierre (JP) de Lutz, from France

[Courtesy Rudy Snel]

Rudy Snel, from Ottawa, Canada

[Courtesy Jean Pierre de Lutz]

A 44-foot sailboat, the *Sean Seamour II*, was JP's pride and joy.

[Courtesy Ben Tye]

Ben Tye, from the United Kingdom

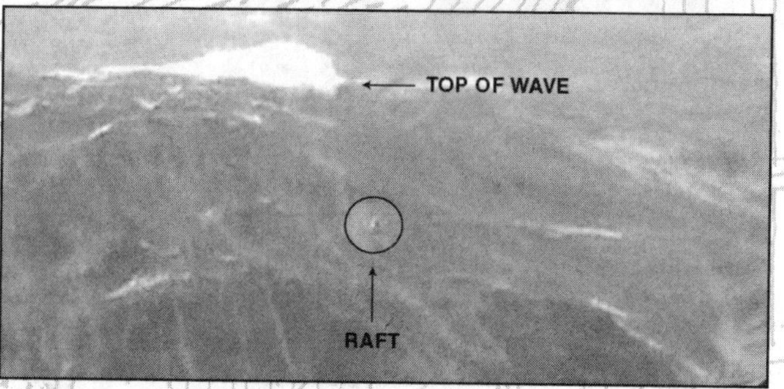

[Courtesy Nevada Smith]

This photo of the life raft that JP, Ben, and Rudy were in was taken by Nevada Smith. The eighty-foot wave makes the life raft look tiny!

[Courtesy U.S. Coast Guard]

The Coast Guard rescue team from left to right: Drew Dazzo, Scott Higgins, Nevada Smith, and Aaron Nelson.

GLOSSARY

BILGE: The lowest point inside a ship.

BOW: The front of the vessel.

CHART TABLE: A table in the cabin of the sailboat where charts can be laid out.

COMPANIONWAY: A raised hatchway on the boat's deck with a ladder leading below to the cabin.

DROGUE: A parachute-like device trailed behind a boat to slow and steady the vessel.

EDDIES: Circular movements of water.

HULL: The watertight structure of the vessel.

HYPOTHERMIA: A medical emergency that occurs when your body loses heat faster than it can produce it, resulting in a dangerously low body temperature.

LOW-PRESSURE SYSTEM: A weather system with lower pressure at its center than the area around it; such a system often produces precipitation.

MICROCLIMATE: The climate of a small area that often differs from the climate around it.

PORT: The left side of the vessel.

ROGUE WAVE: An unusually large, unpredictable ocean wave that appears suddenly and breaks with tremendous force.

SALON: The main social area inside the boat—a place where passengers can relax.

SQUALL LINE: A line of thunderstorms that forms ahead of a cold front.

STARBOARD: The right side of the vessel.

STERN: The back of the vessel.

STROP: A nylon sling connecting the rescue helicopter to the rescue swimmer.

ABOUT THE
AUTHOR AND ILLUSTRATOR

MICHAEL J. TOUGIAS is the author of many true rescue stories for young readers and adults, including the *New York Times*–bestselling adaptation *The Finest Hours: The True Story of a Heroic Sea Rescue*; *A Storm Too Soon: A Remarkable True Survival Story in 80-Foot Seas*; and *Into the Blizzard: Heroism at Sea During the Great Blizzard of 1978*. A frequent lecturer at schools, colleges, and libraries, Tougias divides his time between Massachusetts and Florida.

michaeltougias.com

MARK EDWARD GEYER is the illustrator of three Stephen King novels, including *The Green Mile*, as well as *Blood Communion* by Anne Rice, among other books.

markedwardgeyer.com

If you love the action in the True Rescue illustrated chapter book series and are looking for more in-depth coverage, don't miss these middle-grade true survival stories by Michael J. Tougias.